Why We **Preserve**

The Leelanau Conservancy through the lens of Ken Scott

with essays by Mario Batali, Elizabeth Edwards, Carolyn Faught,

Deborah Wyatt Fellows, Thomas Nelson, Brian Price and Martha Teichner

LEELANAU PRESS

Glen Arbor, MI

Newton Farm Conservation Easement and view of North Manitou Island

Leelanau Conservancy:
Who We Are

Our Mission: Conserving the Land, Water and Scenic Character of Leelanau County

Many would say that there is no more beautiful place on Earth than Leelanau County. Since our founding in 1988 by Ed and Barbara Collins, the Conservancy and its supporters have been devoted to preserving Leelanau's unique natural landscapes, working farms, pristine water quality and sense of community.

As a private non-profit, the Leelanau Conservancy's staff and board partner with thousands of committed donors and volunteers to preserve the most important places in Leelanau. Working with strategic plans and willing landowners, the Leelanau Conservancy protects scenic viewsheds, nourishing wildlife corridors, high quality agricultural land and beautiful lakes and streams. We accept donated land, purchase property and development rights, help families preserve their private lands through conservation easements, and partner with local and national agencies to purchase and accept critical land. Twenty-three natural areas are open to the public to explore, hike and enjoy.

Recognized as one of the nation's leading land trusts, the Leelanau Conservancy was among the first in the country to become accredited through the Land Trust Accreditation Commission. Our first-of-its-kind FarmAbility Program received national attention for its innovative approach to farmland protection. In 2012, the Leelanau Conservancy was selected by the Land Trust Alliance out of more than 1,700 land trusts across the nation to receive its National Land Trust Excellence Award.

We are grateful to serve as the conduit for everyone who loves Leelanau County and wants to preserve this special place for future generations.

DeYoung Natural Area

3

Why We Preserve

THOMAS NELSON

As clear as moonlight on the still waters of the lake, I can recall when Leelanau first sang to me. Can you?

Maybe it was when you were a child, standing with sand-covered toes at the edge of the lapping waves, marveling at a red-orange sun vanishing below Lake Michigan's blue horizon.

Or perhaps it was the sweetness of a ripe cherry or the tang of an apple on a warm afternoon.

It may have been a shiver in hearing the far-off hoot of a great horned owl in the dusky woods.

Or the waft of a bonfire's smoke beneath the stars and the merriment of loving voices close by.

It might have been the shimmer of the northern lights above the cool water and the warmth of someone's hand just before a romantic kiss.

Whenever and whatever it was, in that moment something inside took root. Something within us bloomed. And whether a whisper or joyous shout, it bespoke an awakening:

This is a *Place*.

There are invisible, yet indelible landmarks in the soul — a repository of the touchable Leelanau where our feet or our eyes (or both) have rested. In those spaces our spirit is transported; it soars. Our animal senses become intertwined with these moments, binding them with our pleasures and joys. Oftentimes, we affix them to our very hopes and dreams.

No matter where our daily journeys take us, these distinct sights, scents, and sounds — our sense-memories — ceaselessly flow, wave-like, through us. How miraculous is it that in each of these tangible and intangible places the same name echoes?

Leelanau.

The Leelanau Conservancy has long understood that the landscapes that provide us with both comfort and inspiration can become mere memory if we do not dedicate ourselves to their unbroken existence. (Just ask anyone who has lived for a goodly while.) And not just us, but so too the Community of Life to which we are (as Aldo Leopold so eloquently implored back in 1948) but plain citizens.

Each of us, then, must answer the call to protect what we cherish on our own terms. When it comes to our children, the answer is simple: We make provision. But what of the surf-bent shores, the songbird-trilled footpaths, the verdant fields upon which tomorrow's child will walk?

We *preserve*.

On the following pages, we share with you mere moments in time. Here are images and essays of the many lands and waters we have preserved together through the Leelanau Conservancy. This is our shared legacy. Yes, this *is* a Place! And, the measure of our success is how lovingly this landscape continues to embrace us in our waking thrall, how well we will have secured its blessings so that Leelanau ever sings sweetly to those who come after.

THOMAS NELSON *is Executive Director of the Leelanau Conservancy. "Progress" destroyed the forests and farms of his youth. Tom's quest for a new definition of the term continues through his work at the Conservancy.*

Clay Cliffs Natural Area

Year Preserved: 2013
98 Acres

This stunning natural area created in partnership with Leland Township protects 1,700 feet of shoreline on both Lake Leelanau and Lake Michigan. The sheer clay bluff hosts a rare and fragile ecosystem. Eagles that nest here swoop over the lake in search of prey. The steep, forested slopes are particularly beautiful in the slanting light of dusk, and when the trillium bloom. Clay Cliffs' hardwood forest shelters one of Leelanau's most prolific wildflower sites. An overlook at the bluff top shows off panoramic lake views while a meadow located midway along the trail features sparkling views of Lake Leelanau.

Swanson Preserve

Year Preserved: 2010
83 Acres

The Conservancy had two goals when it purchased this property along M-22 near Sugarloaf. First, to protect 2,000 feet of shoreline on Little Traverse Lake and its two distinct wetland communities critical to wildlife and the health of the lake. Second, to revitalize the adjacent 13-acre iconic farmstead, known as "Sonny's Farm" – home to a beloved farm stand. That land, protected with a conservation easement, has since been sold to a young farmer. Swanson Preserve is one of our most ecologically diverse, with 207 documented plant species, including the rare *Berulaerecta* (cut-leaved water parsnip). A boardwalk winds through cedar forest, over two cold groundwater streams where brook trout spawn, and ends at the lake.

Cedar River Preserve

Year Protected: 1990
440 Acres

This 440-acre undisturbed wetland complex in the Solon Swamp is an ecological powerhouse that protects the health of Lake Leelanau. A number of ecosystems and microhabitats such as fen, shrub scrub and aquatic systems provide a haven for wildlife such as sandhill cranes and rare wetland plants like the carnivorous yellow pitcher plant. The Cedar River is a beautiful, quiet place to kayak or canoe and to enjoy nature at its most diverse. The entire waterway from the mouth of Lake Leelanau to the village of Cedar is protected and open to the public.

Growing Up on the Cedar River

BRIAN PRICE

A wise older man once remarked that he would not want to be young again without wild country to be young in. To my brother and myself, in our mid-teens, the Cedar River was the wildest country on the Leelanau Peninsula. Once our parents were convinced that we could command our 14-foot Alumacraft rowboat with its 18-horse Johnson outboard, we headed for the Cedar River several times each week.

We explored the connecting channels and hummocks of a boggy paradise where our curiosity had free reign. We discovered shallow lakes where the passage of a boat or canoe triggered release of gas bubbles from the mucky sediments, which smelled like rotten eggs. We pulled a canoe over a very long, low beaver dam one spring, relishing the opportunity to paddle connecting channels suddenly accessible along Cedar Run Creek.

An excursion seldom went unrewarded by a new experience or wildlife sighting. It was along the shoals of Cedar Run that we spotted our first bear tracks in Leelanau. Each visit seemed to be rewarded with a purple gallinule or a grebe. If you travel quietly, you are likely to spot a mink or otter, and certainly muskrats. One June morning, we entered the small lake south of Cedar River and found its shallow waters alive with seemingly thousands of spawning dogfish (dogfish!), oblivious to our presence as they were preoccupied in their mating rituals. Brother Bugs and I spotted blue herons, giant pike and an otter feeding on a freshly-killed fish, caught smallmouth bass and dredged the mud banks to screen out "wigglers" (mayfly larvae) that could be sold as bait in Lake Leelanau.

It wouldn't be adventure without the occasional miscalculation. One evening, we decided to float down Cedar Run Creek after supper, pre-positioning a boat near the mouth of the creek where it enters the Cedar River. Little did we know that the creek, as it traversed cedar and tamarack forest, would break into numerous channels, many of them diving under roots to re-emerge 15 feet further downstream. We were dragging rubber rafts over deadfalls and as dark descended, we listened for the sound of water running downstream, our compass home. By the time we reached the Cedar River, it was after midnight. We were cold, wet, and thoroughly beat up.

Over the past 45 years, I've returned to the Cedar River several times each year and in all seasons — often with my own family, the six of us in two large canoes. In a world that changes way too fast, the Cedar River seems relatively oblivious to outside interference. Sure, a few uninvited and unwelcome species have arrived on the scene. But the leatherleaf and bog rosemary, lady's-tresses, pitcher plants and sundews — the entire assemblage of plants uniquely adapted to this cold and waterlogged terrain — is still intact and more than holding its own.

And in recent years, the Cedar River has welcomed back loons and bald eagles and sandhill cranes, all birds that had disappeared for decades, but that clearly belong in this 3,000-acre wetland complex. Cedar and birch trees germinate on the stumps of their predecessors, then spend the next hundred or so years growing to the soaring height of ten feet or so, bonsaied by their inability to extract more than just enough nutrients from the thin gruel of boggy waters needed to survive.

The Cedar River is best experienced by canoe or kayak, silently parting the early morning mist. In so doing you enter as a humble visitor, as you should. We can take some satisfaction from the fact that, perhaps alone among Leelanau's wild places, the Cedar River is least likely to be defiled by the march of civilization into the blank spots on the map.

It will always be there for those who seek to be young again in wild country.

BRIAN PRICE *was the first Executive Director of the Leelanau Conservancy and served for 26 years. He spent his boyhood exploring the natural wonders of the Cedar River with his brother, Larry "Bugs" Price.*

Kehl Lake Natural Area

Year Protected: 1992
279 Acres

Located between the Leelanau State Park and Cathead Bay is a pristine jewel known as Kehl Lake. Three quarters of its shoreline is protected along with 279 acres of wetlands and towering mixed forest. The 100-year-old hemlocks and old-growth white pines which surround the lake provide food and cover for wildlife, particularly birds, and contribute to the lake's exceptional water quality. Its close proximity to Lake Michigan makes Kehl Lake part of a critical flyway for migrating birds and it is also part of an extensive wildlife corridor of protected lands at the tip of the peninsula.

Drawn to Water

CAROLYN FAUGHT

My father and I had what I think of as a loving, yet difficult relationship. He was a complicated man who suffered from mood swings brought on in part by diabetes and died of a heart attack at about the age I am now, 58. When I think of my best times with him, they always involve water.

One of my earliest memories is playing a game we called "Dead Man's Float" out in the lake. None of my siblings or I could hold our breath as long as my dad could. What seemed like minutes would pass and we would beat him about his browned shoulders, yelling, "Dad? Dad? DAD!!!" until at last he would pop his head up with a nonchalant "What?" Which made us scream with laughter, and sigh with relief, too.

He was happiest, I think, out catching perch with my uncle or watching the waves roll in from whatever cottage we happened to rent to escape the heat of downriver Detroit. I recall him standing on the shore, silhouetted by a setting sun. These positive memories are perhaps the reason I am so drawn to water myself, and have made sacrifices to live on this peninsula, surrounded by lapping waves, with pristine inland lakes everywhere you look.

My husband, Dave, and I left big city jobs in Cincinnati in 1987, in large part to be near water. We had purchased a tiny cottage on South Lake Leelanau a few years prior to chucking it all and moving. I recall friends visiting from Ohio, where all the lakes are man-made. Our friends stood waist-deep in ours and were astounded that they could see minnows nibbling at their shins.

The Leelanau Conservancy was founded around this same time and began to protect the dunes and forests, wetlands and streams that flow into or buffer the waters I love. In 2002, I signed on as Communications Director and began to learn about and write about the many ways our work impacts the water quality of Leelanau.

I was told that it was unusual for a land trust to include protecting water in its mission statement. Our founders thought long and hard when they penned ours, which has remained unchanged: "Conserving the Land, Water and Scenic Character of Leelanau County."

Today I live on Omena Bay and I am fortunate to watch the sun come up every morning over sparkling waters. My two boys grew up playing in Freeland Creek across the road and learning to swim on our rocky beach. Unlike me, they take water for granted, because it is all they have ever known. It wasn't until my oldest left for college that he began to understand the power of this place in his life.

To clear my head and get a little exercise on hectic summer work days, I often head to the beach in Leland at lunch time, just a short walk from the office. I leave my shoes in the sand and cool my feet along this breezy shoreline. Whaleback, one of our natural areas, is the majestic landmark I aim for. Many who come to this beach are unaware that the Leelanau Conservancy has protected this geologic wonder and its shoreline, as well as the land by the harbor known as Hall Beach.

Every summer, my husband and I mull the idea of visiting the Grand Canyon or some such place friends tell us we must go. But I just can't bear to leave here during prime beach time. Last summer, as I sat watching the waves pound the shore during a week-long Leelanau "staycation," I read about a study on water and mental health. It talked about how being near water or exercising in water makes everything better, from depression to anxiety to improving the health of a mother and her unborn child.

My dad, of course, knew this intuitively, just as I do. I often wonder how his life would have turned out had he lived where I do, with easy access to clean, fresh water and waves to meditate on and wash our troubles away.

In her role as Leelanau Conservancy Communications Director since 2002, **CAROLYN FAUGHT** *has told the stories of the people who protect their lands and what makes these places so special.*

Whaleback
Natural Area

Year Protected: 1996
40 Acres

This 10,000-year-old glacial wonder near Leland is an iconic landmark that helps to define our unique Lake Michigan coastline. Whaleback's fragile bluff rises 300 feet and can be seen from many vantage points in Leelanau. A viewing platform perched on the edge of this 40-acre natural area offers fantastic views of the Manitou Passage — especially at sunset. Huge oak trees and mature hemlocks create a cathedral-like canopy that shelters bald eagles, and the varied terrain hosts unusual plants like the thimbleberry, which is extremely uncommon in Leelanau.

Whaleback Natural Area

Belanger Creek
Preserve

Year Preserved: 1992
68 Acres

A floristic survey conducted here documented an astounding 170 species, including 30 wildflowers. Come spring, the steep hillsides are full of trillium while sunny marsh marigolds line over a half-mile of protected stream frontage. Three ecosystems host songbirds and raptors, fox and deer. A rich conifer swamp with giant cedars, an old field that is a leftover relic from farming, and a beech-sugar maple-hemlock forest are all present. Six seeps and 11 springs feed into Belanger Creek, and the surrounding 5,600 acres all drain into the stream. Needless to say, protecting Belanger Creek is integral to the health of Grand Traverse Bay.

Lighthouse West
Natural Area

Year Preserved: 2004
45 Acres

Jutting out into Lake Michigan, the tip of the peninsula is a unique landscape. Its cobble shoreline is the last stop for migrating birds to rest and feed before crossing the big water to the Upper Peninsula. With its 642 feet of shoreline, Lighthouse West is a cornerstone at the tip's extensive array of protected lands. Together with the Leelanau State Park and a number of private conservation easements, over 2,000 feet of shoreline and 600-plus acres form a critical wildlife corridor. More than 120 species of birds have been documented at Lighthouse West's diverse and terraced habitat.

Houdek Dunes
Natural Area

Year Preserved: 1998
370 Acres

The story of dune succession plays out here, with dunes of all types present at this dynamic and ecologically diverse natural area. There are forested stabilized dunes as well as dunes that continue to shift and change. There are even blowout dunes — green islands in a sea of sand. Pockets of unusually old white birches are present in the dips of the dunes while pink lady slippers bloom in profusion along sandy trails. This natural area protects much of Houdek Creek, the largest tributary flowing into Lake Leelanau. A diverse array of wetland and upland ecosystems provides abundant wildlife habitat.

DeYoung Natural Area

Year Preserved: 2006
144 Acres

This unique area combines an historic farmstead with a mile of shoreline on Cedar Lake and was preserved in partnership with Elmwood Township. The farmstead dates back to 1870 and serves as a gateway to the county's agricultural corridor. Recreation opportunities abound at this natural area, named for Louis DeYoung, an innovative farmer who died at 104 and dreamed of protecting his land. The Leelanau Trail bisects the land and a Universal Access Trail leads to a lakeside fishing and viewing platform. Upland trails wind past heritage-variety apple trees and over a small stream and offer sweeping views of the farm and Cedar Lake.

DeYoung Natural Area

Hatlem Creek
Preserve

Year Preserved: 2014
40 Acres

Hatlem Creek is the prime tributary flowing into Glen Lake and has long been a priority for protection. The creek and the sensitive wetlands surrounding it provide an important source of fresh water to Glen Lake, and the preserve shelters nearly a half-mile of stream frontage. The federally endangered Michigan monkey flower grows in the area, thriving in wet, mucky soils where cool waters flow. Hatlem Creek is also a wildlife haven; red-shouldered hawks nest in the closed forest canopy and trout and salmon spawn in the stream. A rare bubbling marl spring, 100 square feet in size, is a must-see natural feature.

Krumwiede
Forest
Reserve

Year Preserved: 2007
110 Acres

A glacial moraine formed this high ridge between two scenic wooded and pastoral valleys. It is part of the magnificent hillside that is visible to travelers as they pass through the historic Port Oneida district. This is a high quality working forest where sustainable forestry practices may be observed. Sleeping Bear Dunes National Lakeshore is located less than a mile away. Krumwiede Reserve, in its natural, forested condition, contributes immensely to the ecological integrity of both Leelanau County and the National Lakeshore.

Loving Tribute

MARTHA TEICHNER

It was the wildness of the land my parents learned to love when they found themselves in northern Michigan after World War II. Newly married, they bought the only home they would ever own together. They called it Deer Trail Cottage, the old white house my mother spotted, boarded up and neglected, on the road around Lime Lake. With it came 40 acres of woods and lake frontage. A lumber baron named Fisher built it just after the Civil War. He had a saw mill through the trees on the water.

I was nine when my father died in 1957. My mother had to sell the house and our land on Lime Lake. It was like tearing her heart out, having to leave this beautiful place. She never came back. It was too painful. Our lives were defined by longing for what was gone.

Our only foothold left was the 20 acres next door that my grandmother had bought before she too died. One by one, my uncles and aunts stopped paying their share of the taxes, so in the end, I inherited the property from my mother.

Not long before her death in 1992, I asked her to relate the most romantic thing she and my father used to do together. In all of my childhood memories of them, they were young and strong and wonderful. She sat for a very long time and then said, very quietly, very simply, "In the summer, when the moon was full, sometimes at night when you were in bed asleep, we would go down to Lime Lake. We would push out the raft and swim in the moonlight."

It had been her secret all those years, a special treasure, so precious she only allowed me a glimpse of it as she, herself, was dying. In my mind I could see the depth of the woods closing in around them as they disappeared down the path to the lake, hear their breathing, their footsteps on the soft earth. And then they would slip into the water, cool and smooth and silvery in the night stillness, rippling softly. Caressed by moonlight, the sky alive with stars, they swam, bathed in the mystery of being utterly alone there.

They are buried together in Leland, under a big pine tree.

Twice, luck and timing have positioned me, with the help of the Leelanau Conservancy, to honor my parents and their love — my love, too — for this special place.

In 1996, I attended an event honoring my father, Hans ("Peppi") Teichner. Funny, colorful, his thick Bavarian accent a local joke, he brought skiing to this part of Michigan. My hosts drove me around Lime Lake, past the house and the property I now owned. Through my trees, I saw bulldozers clearing land for a golf course at Sugar Loaf.

I have nothing against golf courses, but those growling, tooting, beeping machines represented a future that scared me. Someone in the car suggested calling the Leelanau Conservancy, of which I'd never heard, and I donated my precious 20 acres. Those acres began the Teichner Preserve.

By chance in 2005, I met the Blakelys, the current owners of Deer Trail Cottage. All the land around it had been sold off long ago. A speculator, they said with resignation, was days away from filling in the wetlands and building in the woods. I felt sick.

Again, I called the Conservancy. Refinancing my apartment in New York., borrowing $200,000 just to give away was frightening, but I did it with joy and no hesitation. The Conservancy put up another $200,000. Together, we stopped the development, and the Teichner Preserve doubled in size.

If I had done nothing to preserve land that is as much a part of me as my name, my face, my blood, I doubt I could have lived with myself.

I'd like to think that my parents, under their pine tree in the Leland cemetery, somehow conspired to put me in the right place at the right time. This beautiful spot was the backdrop for their life together. It gave them great happiness. It gave me an important part of my identity, my sense of the world; it was time to repay that favor. I am proud to tell people that if I've done one good thing in my life, helping to keep a part of Lime Lake wild is it.

Teichner Preserve

Year Protected: 1996
41 Acres

The preserve includes 200 feet of natural shore-line on Lime Lake, framed by mature cedars. A boardwalk traverses the fragile forested wetland and offers views of a vibrant and healthy ecosystem. Come spring, the sounds of birds, frogs and trees creaking in the wind fill the air, while wildflowers put on a spectacular show along the boardwalk. Tamarack, elm and a giant surviving American chestnut tree also call Teichner home.

CBS NEWS Sunday Morning correspondent **MARTHA TEICHNER** *spent her early childhood on this land along Lime Lake and was instrumental in creating the preserve that honors her parents, Hans and Miriam Teichner.*

Crystal River

Year Preserved: 2004
104 Acres

This beloved river in Glen Arbor meanders between Glen Lake and Lake Michigan and was preserved in partnership with Sleeping Bear Dunes National Lakeshore. Over a several year period, the Conservancy worked with multiple parties and secured funding to preserve this sensitive "dune and swale" topography once slated to become a golf course. The preserved land includes a mile of shoreline that the U.S. Fish and Wildlife Service recognized as "globally rare habitat." In 2003, the Conservancy also purchased the fragile 7-acre oxbow portion of the river and transferred it to Glen Arbor Township protected with a conservation easement.

Forever Crystal

ELIZABETH EDWARDS

On September 27, 2009, I stood on the banks and snapped a photo of my 12-year-old son standing ankle-deep in the sun-ringed, sandy-bottomed Crystal River. He's boy-handling a monster salmon he'd just caught, one hand on the tail, the other cradling the head with its powerful maw agape, teeth sharp as finishing nails.

We'd done this routine before: I'd drop Keef off near the river mouth, where it empties into Lake Michigan's Sleeping Bear Bay, with his rod, a net, a stringer, a cell phone in his pocket and Teva sandals on his feet. He'd walk up the river toward Glen Arbor, casting along the way. The deal was that he'd call when he caught a fish and I'd meet him with a bucket. Sometimes he'd walk as many as three miles through the knee-deep water, peering for salmon in every dark, tree root-lined hole.

The day I took the photo was one of those northern Michigan September afternoons when several cool nights have bleached the atmosphere of summer's dust and fuss. Everything in the frame reflects the sun: Keef's orange basketball shorts, his green striped polo shirt, his smile, the seaweed green scales of Mrs. Salmon. And the water — those sun circles in the foreground give way to ripples and dances of light behind him.

It is a wonderful photo, but any image is better when you know the context. And I understand the crooks and wending of this 6.3-mile river better than I do plenty of other things.

I know where the muskrat and otter live. Where the deer cross, where the pike school up, where the smallmouth bass hang out, where the craggy old snapping turtles burrow into the mud and where to pick wild blueberries. I canoed it the first time at age 2, that memory still bright as the bulky flame-orange life jacket that made me sit like a tiny Buddha. My father in the stern easily finessing the gentle current, mom in the bow sporting her Jackie Kennedy sunglasses and my older brother flopping behind me like a netted fish. I had just gotten the courage to peer over the aluminum side, straight through the shallow, glass-clear water to the colorful gravel (orange, black, white, green-gold — leftovers from the melt of ancient glaciers that carved this river valley) when he gave the canoe a pronounced list starboard. I froze. "Help, help, help—help," my mom still quotes me saying all these years later.

So much followed over the years — dog paddling in the swimming hole across the street from the old Shell gas station where the attendants handed out fabulously huge truck inner tubes for floating in, shooting the culverts under County Road 675. Canoeing the Crystal with my young family and our best friends on so many Sunday afternoons, children and snacks packed to the gunwales.

Rivers have their way, calling you back (Mrs. Salmon herself being more than metaphorical here), washing perfect moments into memories, cleansing others away. And so it is — that although I was deeply involved — I can barely recall now the tempest that led to the Leelanau Conservancy stepping in to forever preserve many acres of the loveliest wetlands along this river, land that had been targeted for a golf course. It was a tangled and sensitive affair knit of local, state and federal governments, roiling passions and millions of dollars.

All of those things drifted through my mind the day I leaned out over the banks of the Crystal to take my son's photo. But mostly I thought this, and I thought it very clearly: How much better the world would be if every young boy had a crystal-clear river he could cast his way down to catch a fish, bring it home, filet it and cook it up himself for dinner.

ELIZABETH (LISSA) EDWARDS *is Managing Editor of* Traverse, Northern Michigan's Magazine. *She raised her family in Glen Arbor, and was part of the effort to ensure the protection of the Crystal River she loves.*

Narrows
Natural Area

Year Preserved: 2001
71 Acres

This iconic wetland that joins North and South Lake Leelanau is perhaps one of Leelanau's most viewed landmarks, seen from the bridge on M-204 and by boaters who traverse the channel. The Conservancy recognized early on that The Narrows, one of the finest remaining emergent wetland complexes on the lake, must be protected. Over 10 years, three important parcels were acquired, including the Leugers Preserve. The most recent acquisition was the site of a proposed 80-slip marina. Precious, intact wetlands and 2,350 feet of shoreline provide important fish and bird habitat. Raptors nest here and numerous grasses provide abundant wildlife habitat.

Adventures on The Narrows

DEBORAH WYATT FELLOWS

There is a small, magic passage connecting North and South Lake Lee-lanau that is referred to as simply The Narrows. This ribbon of water winds through grass-filled areas that offer safe harbor to families of swans and into narrow river passages that flow past wide, wooden docks topped by young kids fishing for bluegill or boats tied by folks who have walked up into the village of Lake Leelanau.

The Narrows is exquisitely beautiful any month of the year. I'm often hard-pressed to decide: Is my favorite our first drift through in the spring, when baby ducks and swans follow noble parents through the grasses that sway in that first warm, fresh air? Or the vision of The Narrows in winter, framed in snowy shores, where crystal branches gleam like intricate mosaics over water that still moves purposefully from one frozen lake to the other? The bridge over The Narrows, just as you enter the village, offers endless opportunities for stunning photographs no matter which direction you face — perfect scenes that frame old boathouses against meadow grasses, birch and evergreens, a natural landscape forever-protected by the Conservancy.

The Narrows is a no-wake zone. That means, in the height of summer, the speed (and hilarity often accompanying a day on the lakes) is left behind at the bobbing marker indicating the start of The Narrows. The world slows down perceptively and occupants settle in to glide through the river, shared with kayakers and paddle boarders, accepting and engaging with the temporary sense of calm.

When our kids were little, entering The Narrows was a time of curling up in the front of the boat to gaze up at the clouds or gathering at the bow to peer into the marshes for signs of fish or turtles. Rarely does The Narrows disappoint in revealing at least one treasure: a gaggle of babies and their mom, a snapping turtle resting upon a branch, a school of fish darting among gracefully-swaying lily pads, their blossoms creating watercolors no matter where you look.

One of the first solo adventures each of our kids was allowed to embark upon was an excursion into The Narrows before and after the busy days of summer. Armed with food enough to last for a week, they would head out in our small fishing boat, the bony shoulders of youngsters just growing into their bodies, dwarfed in their life jackets. They would go ostensibly to fish, but they never returned with any catch, leading us to think they were much more likely to have been on a pirate adventure, stalking imaginary foes than baiting their hooks.

The first trip made was when our oldest son, Ben, was 10, and as such, was the first to be allowed to take out a new little fishing skiff. He had wanted an electronic, hand-held game console called a Gameboy for his 10th birthday. He was gracious in his clear disappointment, but excited the first time he, his friend, Eric, and our second son, Peter, packed the boat for a fishing adventure to be found in The Narrows on an early morning in June. We half expected them to be back within the hour and so found ourselves looking toward The Narrows' opening several hours later with the keen sense of worry that comes when kids and water are involved.

Sometime late morning, here they came, returning our waves with grins, Peter literally bouncing in the front of the boat. All talking at once, we heard of all they'd seen and of calamity when the rope got caught in the motor and Peter paddled to shore to seek assistance that came from a lovely man who untangled the rope and sent them on their way. The food was gone. There were no fish. In their place stood three young boys who had headed into a natural place of beauty and intrigue and found adventure, joy and the priceless gift of confidence.

Two nights later, Ben took the boat out for a short spin just as the sun was setting, and that magic light gleamed upon his hair, his shoulders and upon the name his dad and he had stenciled on the back of the boat: *Gameboy*.

For that, and much more, I will forever be grateful to and enchanted by The Narrows.

DEBORAH WYATT FELLOWS *is a past Conservancy board member and founder of* Traverse, Northern Michigan's Magazine. *She and her husband, Neal, have raised their family near the Lake Leelanau Narrows.*

Chippewa Run Natural Area

Year Preserved: 2000
110 Acres

Years ago when a beloved parcel on the outskirts of Empire went up for sale, the community and the Leelanau Conservancy rallied to preserve this scenic buffer which protects the village's small town character. The land includes four separate ecosystems, and features a much-loved beaver pond and stream where brook trout spawn and where blue flag iris and cardinal flower grow in abundance. Chippewa Run is a birding paradise where dozens of species visit or live, including green herons which nest in the pine tree grove that was planted in 1953.

Jeff Lamont Preserve

Year Preserved: 2008
40 Acres

Friends and family of Jeff Lamont came together to preserve this land in his memory. Jeff, whose family owns a cottage nearby, died of cancer just after his 21st birthday. He adored Leelanau, and so coming together to create the preserve helped those who loved Jeff to remember and to heal. This forested wetland at the tip of the peninsula features a magical trail through maple, hemlock and beech, dotted with pink lady slippers and other wildflowers come spring. A dense wetland thicket with ferns and 6-foot-tall cattails provides important wildlife habitat. Canopy birds such as the scarlet tanager sing from the treetops.

Gull Island
Preserve

Year Protected: 1995
7 Acres

Gull Island, off the coast of Northport, is preserved as a sanctuary for thousands of herring gulls, whose populations have declined in the last 25 years. Don't confuse these birds with ubiquitous "sea gulls," which herring gulls help to keep in check. The birds, which nest on Gull Island from April to June, mate for life and will abandon their nests at the slightest disturbance. In a long term study here, banded birds have returned for as many as 24 years. A crumbling stone cottage on the island was once inhabited and is a unique part of Northport's history.

Casier Farm Conservation Easement

Private Lands Protected by **Conservation Easements**

When most people think about the Leelanau Conservancy, spectacular natural areas like Clay Cliffs or Kehl Lake come to mind. These places are open to the public to hike and enjoy. But over two thirds of the lands we have protected are private lands, owned by families who cherish these special places and want to see them preserved for generations to come.

The Conservancy has worked with over 150 willing landowners to preserve family farms and treasured woodlands, significant wetlands and sweeping shorelines with conservation easements. We look upon all of our conservation easement landowners as heroes, and our guiding principle is that each land transaction must be good for the land and the people involved. These private lands stay on the tax rolls and are not open to the public, but the public benefits enormously from their protection. The view-sheds, wildlife habitat and water quality that conservation easements help to protect are a significant part of our legacy to date.

Olsen Farm Conservation Easement

Sedlacek Farm Conservation Easement

Paradise Found

MARIO BATALI

By pure luck I landed in Leelanau in 1995, following my wife Susi and a handful of her college friends. When I am here, I wake up before the sun just about every summer morning and around the holidays in November and December. The sun peeks out over Charlevoix and illuminates the Old Mission Peninsula like a candle in a hall of mirrors. The wind changes direction but often blows in firm from the north — or lightly, straight over the house, headed east with the clouds as they meander toward Elk Rapids. It smells sometimes like a blend of green tea tempered with the salt-free, nearly marine fragrance of the big lake, resting quietly off of Cathead Bay on the other side of the peninsula.

To walk our beach north of Northport is to observe the constant ebb and flow of life force on Lake Michigan. Most often sweet, clear, and calm, sometimes we find movement in a blanket of mayflies, left over from a summer hatch near the shore. Sometimes the waves churn in big, reminding us of a storm coming out of Sault St. Marie or Escanaba. Sometimes, this great freshwater sea is the placid flat nap of what can only be described as Caribbean. The lake is ever-changing and evolving in ways we learn more about every year.

Before I knew Leelanau, I first moved to Italy in 1989 and landed in the tiny town of Borgo Capanne on the River Reno, the border between Toscana and Emila Romagna. I found a place filled with the optimism of good farming, the delicious seasonality of the fields, their fruits and vegetables, and the poetic rhythm of life. I uncovered joy in the creation of wine and cheese, ale and cider — in the sustainable partnership between man and soil and the spirit of the two intertwined. Even in a place settled since the Etruscans in 1000 B.C., I found kilometer after kilometer of undeveloped land, much planted, but nary a building or commercial activity — just a vast, unspoiled canvas expressing its own pure beauty and majesty.

After three years in my Italian slice of heaven, I left Borgo Capanne to pursue my fortune in America, but I never forgot the powerful joy of life in a place that is engaged equally in commerce and stewardship of the land — a place that understands its sacred responsibility to protect the land for the next generation and the one after that.

During Leelanau County summers I ride my bike to the lighthouse, to town, or down what we call "Barb's Trail" from Suttons Bay to Black Star Winery to grab a pizza and a glass of local Cab Franc. I ride from the Sleeping Bear Dunes to Glen Arbor to scarf down ice cream at The Pine Cone, especially during cherry harvest. For shits and giggles we will stop by Tandem Ciders to show off the crazy little pub that reminds me of the Slug and Lettuce in Islington, U.K., then hit L. Mawby Vineyards' tasting room to celebrate "Mawby-ness" in a way that feels more like an agriturismo in Trentino than anything I have experienced in California, or anywhere else in the U.S.

On Thursdays, the farmers' market is in Leland; on Fridays, it's in Northport; on Saturdays, Suttons Bay. I shop like an Italian nonna, checking each stand for the best favas, lettuces, peas, ramps, sugar snaps, Kirby cukes, beets, chard, and eventually raspberries, then cherries, apricots, and peaches. There is goat cheese from Idyll Farms and raclette from Leelanau Cheese Company. There's a guy who makes spanakopita and a lady who makes my favorite jam and pickles. There are farmers we love like Bare Knuckle; TLC Tomatoes' little greenhouse shop; the Ugly Tomato Farmstand; and dozens of opportunities to "Pick Your Own," paired with road-side stands for those who are hungry right now.

There are miles and miles of rolling hills covered with stone fruit and grape vines — often with breathtaking views of the water or Manitou Islands or the Old Mission Peninsula, itself a nascent poetic landscape of viticulture and gastronomy.

Over the past two decades, this culture of good eating in Leelanau has been discovered by many others, through its restaurants and a food scene rivaling hot spots like Charleston, South Carolina; Austin, Texas; or even Napa Valley. A dozen national publications have featured this place as "paradise found."

Consider my temporary home in Borgo, Capanne. After inhabitation by the Etruscans, followed by the Romans, the Goths, the Longobards, then Charlemagne, the Medicis, the Hapsburgs, the Bourbons, Napoleon, dukes, duchesses, kings, queens, popes, an industrial revolution, two world wars, and a great and lesser Depression — it is still untouched. Despite thousands of opportunities to commercialize and exploit the natural beauty with development and infrastructure, the Italians remain unmoved and undeveloped, steadfast in their understanding of the timeless and remarkable value of untouched land.

I remain certain that those of us who love Leelanau can learn a lesson from the people of central Italy and envision this beautiful landscape preserved for decades to come.

Chef **MARIO BATALI** *and his wife, Susi, are among the Leelanau Conservancy's strongest supporters of farmland preservation and are also champions of the local foods movement.*

Spinniken Farm Conservation Easement

FEATURED NATURAL AREAS

1 Belanger Creek Preserve
2 Cedar River Preserve
3 Chippewa Run Natural Area
4 Clay Cliffs Natural Area
5 Crystal River
6 DeYoung Natural Area
7 Gull Island Preserve
8 Hatlem Creek Preserve
9 Houdek Dunes Natural Area
10 Kehl Lake Natural Area
11 Krumwiede Forest Reserve
12 Jeff Lamont Preserve
13 Lighthouse West Natural Area
14 Narrows Natural Area
15 Swanson Preserve
16 Teichner Preserve
17 Whaleback Natural Area

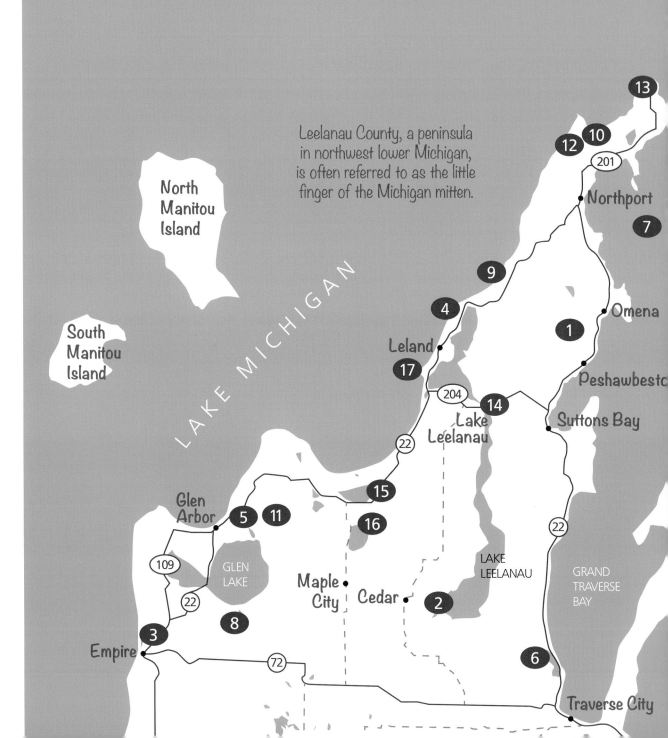

Leelanau County, a peninsula in northwest lower Michigan, is often referred to as the little finger of the Michigan mitten.

KEN SCOTT has become a local icon by photographing Leelanau County for 30 years. Having produced six books and contributed to the back page of the *Leelanau Enterprise* since 2005, he has a wide following and actively engages his audience to experience nature through his lens. He was commissioned by the Leelanau Conservancy to produce images of their preserved land which resulted in an exhibit at the Dennos Museum and this book. Scott feels it has been a unique opportunity to capture images of these protected Conservancy properties, some of which are not open to the public. He is represented by several galleries and bookstores and can be found at KenScottPhotography.com, Facebook, YouTube and Flickr.

Photography ©2015 Ken Scott

Published by Leelanau Press in conjunction with the Leelanau Conservancy

Leelanau Conservancy
105 N. First Street, Box 1007, Leland, Michigan 49654
231-256-9665, LeelanauConservancy.org

Leelanau Press wishes to thank Angela Saxon of Saxon Design Inc, book design; Lisa Jensen, copy editing; and Sylvia Duncan, proofreading.

Leelanau Press
Box 115, Glen Arbor, Michigan 49636
231-334-4395, siepker@aol.com, LeelanauPress.com

Printed in Canada by Friesens

Library of Congress Control Number: 2015935007

ISBN 9780974206820

10 9 8 7 6 5 4 3 2 1